SHARKS

WHALE SHARKS

by Julie Murray

Cody Koala
An Imprint of Pop!
popbooksonline.com

Hello! My name is Cody Koala

This book is filled with videos, puzzles, games, and more! Scan the QR codes* while you read, or visit the website below to make this book pop.

popbooksonline.com/whale-shark

*Scanning QR codes requires a web-enabled smart device with a QR code reader app and a camera.

abdobooks.com
Published by Pop!, a division of ABDO, PO Box 398166, Minneapolis, Minnesota 55439.

Printed in the United States of America, North Mankato, Minnesota.

052023
082023

THIS BOOK CONTAINS RECYCLED MATERIALS

Cover Photo: Blue Planet Archive
Interior Photos: Shutterstock Images; Getty Images; Blue Planet Archive
Editor: Elizabeth Andrews; Grace Hansen
Series Designer: Victoria Bates

Library of Congress Control Number: 2022950505

Publisher's Cataloging-in-Publication Data
Names: Murray, Julie, author.
Title: Whale sharks / by Julie Murray
Description: Minneapolis, Minnesota : Pop!, 2024 | Series: Sharks | Includes online resources and index
Identifiers: ISBN 9781098244262 (lib. bdg.) | ISBN 9781098244965 (ebook)
Subjects: LCSH: Whale shark--Juvenile literature. | Sharks--Juvenile literature. | Sharks--Behavior--Juvenile literature. | Marine fishes--Behavior--Juvenile literature.
Classification: DDC 598.47--dc23

Table of Contents

Chapter 1

Largest Sharks

Whale sharks are the largest type of shark. They are also the biggest fish in the world! They can be 40 feet (12m) long and weigh up to 47,000 pounds (21,320kg).

Watch a video here!

Whale sharks mainly live near the **equator** in warm ocean waters around the world.

They like to be in open water far from shore. They are often spotted near the water's surface.

Whale sharks live and swim alone. They may gather with others to feed for a short time.

They **migrate** thousands of miles each year in search of food.

Chapter 2

Unique Pattern

Whale sharks are brown, gray, or blue in color. Their bellies are white. They have seven fins that help move them through the water.

Whale sharks are slow swimmers. They only swim about three miles per hour (5kph).

Learn more here!

Whale sharks have spots and stripes. Each whale shark has a **unique** pattern. Like human fingerprints, no two whale sharks have the same pattern.

Whale sharks have thick skin. It can be six inches (15.4cm) thick!

Chapter 3

Filter Feeder

The whale shark has a giant mouth. It can open five feet (1.5m) wide! Inside the mouth are thousands of tiny teeth and **filter pads**. The shark's teeth are almost useless.

Explore links here!

Whale sharks take in and push out around 10,000 gallons (38,000L) of water every day to eat enough food.

A whale shark eats by opening its huge mouth. Water and food rush in. The water is able to pass out through the pads, but the food stays trapped. The giant shark eats tiny **plankton** and small fish.

Chapter 4

Life of a Whale Shark

Whale sharks begin life as eggs in their mother's body. They hatch inside their mother and are born live. Females can give birth to up to 300 young at a time.

Complete an activity here!

After birth, young whale sharks are on their own. It will be about 25 years before they can have their own young. Whale sharks can live for 100 years or more!

Making Connections

Text-to-Self

Whale sharks are filter feeders. How is that different than how you eat your food?

Text-to-Text

Can you name one fact that you learned about whale sharks in this book or a different book you have read?

Text-to-World

Whale sharks are the largest shark in the world. Can you name some other large animals?

Glossary

equator – an imaginary line around the middle of the Earth, halfway between the North and South Poles.

filter pad – one of several pads throughout a whale shark's mouth. The pad has many millimeter-wide openings that allow water to pass through while capturing food.

migrate – to move from one place to another, often to find food and have young.

plankton – tiny animals and plants that float in a body of water.

unique – being the only one of its kind.

Index

Online Resources

popbooksonline.com

Thanks for reading this Cody Koala book!

This book is filled with videos, puzzles, games, and more! Scan the QR codes* while you read, or visit the website below to make this book pop.

popbooksonline.com/whale-shark

*Scanning QR codes requires a web-enabled smart device with a QR code reader app and a camera.